Power Assisted Magnet Motor

THEORY OF OPERATION:

CONVERTING MAGNETIC FORCE INTO POWER

Author: Donald Sitler

Printed in the United States of America

ISBN:

ISBN-13: 978-1530935086
ISBN-10: 1530935083
Cover Design: Don Sitler
Interior Design: Don Sitler
Library of Congress Control Number: 2016906096
CreateSpace Independent Publishing Platform, North Charleston, SC

DEDICATION

This book is dedicated to my children Michael and Tiffany.
Also to my granddaughters, Devynn, Isabella (Gracie) and Anna.

It is also dedicated to the addition of science and inventors who have worked with permanent magnets, and found how many ways that will not work in a motor and have experienced frustration. Be of good cheer, for the answer is found in this book.

I would like to thank my wife Cindy and my parents for their encouragement through difficult times to persevere. Most of all, I would like to humbly thank God, and acknowledge the King of Kings, the Lord Jesus Christ by giving Him glory, for giving me the spirit of wisdom and knowledge of witty inventions through His Holy Spirit.

CONTENTS

ACKNOWLEDGMENTS

I would like to acknowledge and give special thanks to Magnetic Shield Corporation for their part as a supplier and their professionalism and kindness. I would like to acknowledge inventors and experimenters who have put in long hours and study to try to understand magnetism and electronics. I began to find out how permanent magnets can work together successfully, to perform work, in the form of a motor. This mission has inspired me to self study Physics, and increase learning for future magnetic projects. This project has also led me to start my business web site: Dynamaticmotors.com and eventually led to the writing of this book, to tell others about a new form of motor. The Power Assisted Magnet Motor possesses special advantages enabling high efficiency operation.

CHAPTER 1: PERMANENT MAGNETS

One weekend, I was working on my car in the garage, and Dad walked up to see how I was doing. I asked him a question about why permanent magnets weren't being used in motors, since they produce a magnetic force without power input. He replied, that the biggest problem with using magnets is that the magnets can't be turned off, they remain "on" all the time. A motor will turn until the magnets lock, then it will not rotate unless you can turn the locking magnet off to release the magnetic lock. Dad went on to explain that electro-magnets are used instead of magnets in motors, since they are powered, they can be turned on and off at the proper time, enabling the motor to spin.

POWER ASSISTED MAGNET MOTOR

Many years passed after that conversation, but the conversation remained in my mind as an unsolved equation, that keeps coming back into your thinking from time to time. I have seen several attempts by people to make a pure magnet motor, using many magnets. The motor has weak rotation or incomplete rotation, but then ending in a magnetic lock or a bounce off of a repel magnetic field to stop rotation. I begin to think, how can a motor rotate with magnets without attraction lock or repel collisions. I started to formulate a way to solve that equation from years ago, and demonstrate it in a model motor.

I began to experiment with magnets in my hands to learn how far the magnetic fields extended and what shape they were. I learned that most magnets, even with different shapes, seemed to have a spherical magnetic field. I used the repel field of the magnets to measure the distance that the magnetic fields would interact. I used this method of field measurement of the magnets to establish the proper distance to space magnets in a rotor. Next, a configuration of magnets would have to be established, and if possible the strongest configuration should be used. I experimented by using weak magnets, with stronger ones following in attraction, to move the rotor around. However, at the strongest magnet, at the end of the string, it would stop with a strong magnetic lock. I knew if an electro-magnet was used to break the lock, it would have to be at least as strong as the strongest magnet. In conclusion, I decided that the motor should not have a coil directly in the motor for efficiency, in order to break the magnetic lock.

I built a rotor with a metal band out of **NETIC®** metal, which the intended use was for magnetic shielding, meaning it was a metal that could "conduct" heavy magnetic flows. I used this metal band to attach and house the magnets in the rotor. I decided that the strongest combination of magnets would be from repel to attract. I alternated the rotor magnets while using the spacing determined in earlier experiments. Now I had a rotor on bearings with two magnets. I used an eight inch diameter metal **NETIC®** band, which held the rotor magnets with the two magnets spaced at the half-way point, to remain in balance during rotation. I knew with alternating polarities of rotor magnets, that if I started by holding a magnet in hand as the field magnet, the rotor magnet should repel away from the field magnet, then attract to the second rotor magnet. This achieved half of a revolution ending with an attract magnetic lock. This completed half a revolution without much input. I noticed that while starting from the repel field, that if the field magnet was stationary, but in the repel field of the rotor magnet, it would push away with moderate force. However, if I quickly moved the field magnet in my hand into position to repel the rotor magnet, the repel force was increased and rotor speed was increased, due to a higher level of energy input to start the motor. It seemed that if the repel field suddenly appeared in repel to the rotor magnet, the repel force for rotation was greater. While experimenting with this new "model", I found that if I moved my hand with the field magnet up and down in

specific time with the rotor magnets, that I could switch

fields, using both sides of the magnets in the rotor from

repel to attract again, completing one revolution of the rotor.

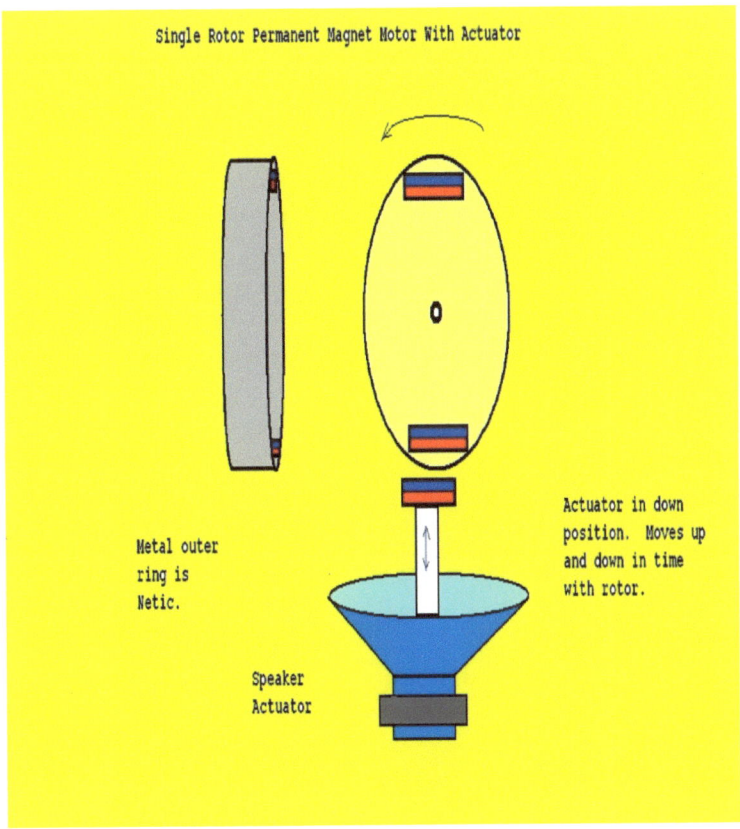

A simple diagram of a single rotor Power Assisted Magnet Motor.

CHAPTER 2: MAGNETIC PROPULSION

Magnet arrangements have a lot to do with magnetic propulsion and the efficiency and use of permanent magnets in a motor. One of the main issues in a magnet motor is to ensure that there are no or very little magnetic collisions. A magnetic collision occurs when two repel fields collide with one another, or in attraction a magnetic lock occurs during rotation of a magnet motor. One collision will sap the power from a magnet motor and possibly damage the pinning material that holds the magnetic particles in place in a magnet, resulting in the weakening of the magnetic field of the magnet.

POWER ASSISTED MAGNET MOTOR

The magnets used in the Power Assisted Permanent Magnet
Motor are N 38 grade strength, and are one inch wide, by two
inches long, by half inch thick. They are rectangular in shape.
The shape and pole location is important in motor
performance. The pole is located on each flat side of the
magnet. The pole is located or concentrated in the center of
the magnet. It is important to understand this, as when
magnets attract, they continue to pull toward each other until
the center of the magnet is found. Then due to built up
momentum, the magnets will pass each other momentarily
before pulling back to lock. This action I call the "magnetic
window" and is an important discussion in the next chapter.
The next illustration is an example of the magnetic interaction
between three permanent magnets for motor propulsion,
using the strongest combination of magnetic fields and the
fewest magnets for efficiency and cost reduction.

POWER ASSISTED MAGNET MOTOR

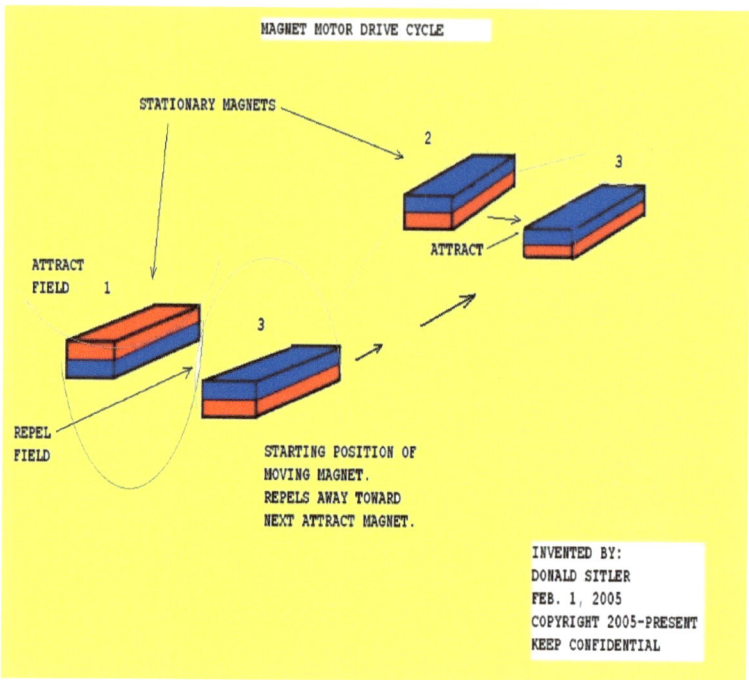

Magnet propulsion illustration with field magnet 3 in motion.

The picture above shows the starting position of the repel field magnet (3), with repel fields of magnets (1) and (3) portrayed as three dimensional, spherical bubble shaped fields, pushing away from each other. This produces the initial propulsion in this magnetic interaction. In the magnet motor, magnets (1) and (2) are mounted on the rotor, while magnet (3) is the field magnet that is mounted on the

8

actuator, that moves up and down in time with the rotor magnets.

With magnet (3) starting in the repel position, interacting with magnet (1), magnet (3) in this example moves away from magnet (1), until it leaves the repel field of magnet (1), and then starts entering the attraction field of magnet (2). Magnet (3) continues its travel towards magnet (2) in attraction, completing one-half of a rotation in a magnet motor. So far, we have not needed to use the power assist portion (the actuator), to assist rotation. The magnets are the only propulsion in the motor at this point of rotation or magnetic interaction. This accomplishes the first half-cycle of operation.

The second half-cycle operation is identical to the first half-cycle operation. The exception is that we need to accomplish a magnetic field switch via the actuator, which we will cover in the next chapter. This field switch will allow us to use the other side of the rotor magnets (3) and (2) to accomplish the second half-cycle of rotation of the motor. This returns the rotor back to the interaction between field magnet (3) and rotor magnet (1). With another magnetic field switch, the rotor will be ready to repeat these two cycles of magnetic interactions, to continue rotation. In effect, this motor links these two repel/attraction cycles together similar to a stepper motor. Most of the magnetic propulsion is provided by the permanent magnets interaction with each other in both cycles of motor operation.

In summary, we have chosen to use the strongest magnet configuration of repel to attract for propulsion, and we use the field switch to be able to use the other side of the magnet in the rotor, for efficiency and close field switching. This enhances motor speed by using the side of the magnet, allowing quick magnetic field switching. We did not choose to use a gradient or slope/spiral of magnets, due to poor torque and the number of magnets required, which would result in increased motor costs and inefficient use of magnetic

fields for propulsion. By placing magnets (1&2) in the rotor

(instead of a straight line arrangement), the field magnet

remains as the magnetic point of interaction. This allows the

magnets mounted in the rotor to rotate, rather than the field

magnet, which is mounted upon the actuator in a stationary

role.

CHAPTER 3: SWITCHING FIELDS

This is the heart of the Power Assisted Magnet Motor. Here is where we connect the first magnetic cycle to the second magnetic cycle for complete and continued rotation of the motor. To start, the magnetic window which is the velocity that is accumulated within a magnetic cycle, causes the rotor magnet to go past the field magnet while in the final attraction phase of a magnetic cycle. The swing past the lock point of the two magnets in attraction creates a "magnetic window", to accomplish the magnetic field switch. There are a number of reasons that the magnetic window is used, rather than just switching at the time of magnetic lock or before. First of all, if the switch occurs at the time of lock or before, the motor rotation velocity will be lost, and the magnet interaction is at the strongest point, requiring maximum energy input to accomplish the magnetic field switch to the next magnetic cycle by the actuator. Additionally, since the magnetic field at the magnet is a sphere, if the switch occurred at lock, the repel field force would be to the side, rather than in the direction of rotation of the rotor. Therefore, it is determined by this study of the magnetic window, that the field switch must occur at a time where the magnets are slightly past each other. This magnetic switching occurs at such a time and in an area where the magnetic interaction of repel fields will occur, so as to boost the rotation of the rotor with the kinetic energy input of the actuator. The repel fields with this timing, cause the force from the field switch to aid the rotation of the rotor. This causes most of the input energy expended in the actuator to switch magnetic fields, to inject this energy into the rotation of the rotor, while causing the next magnetic cycle to occur. It all works together and increases the motor's overall efficiency. In this next illustration, a demonstration of the

magnetic field switch is shown.

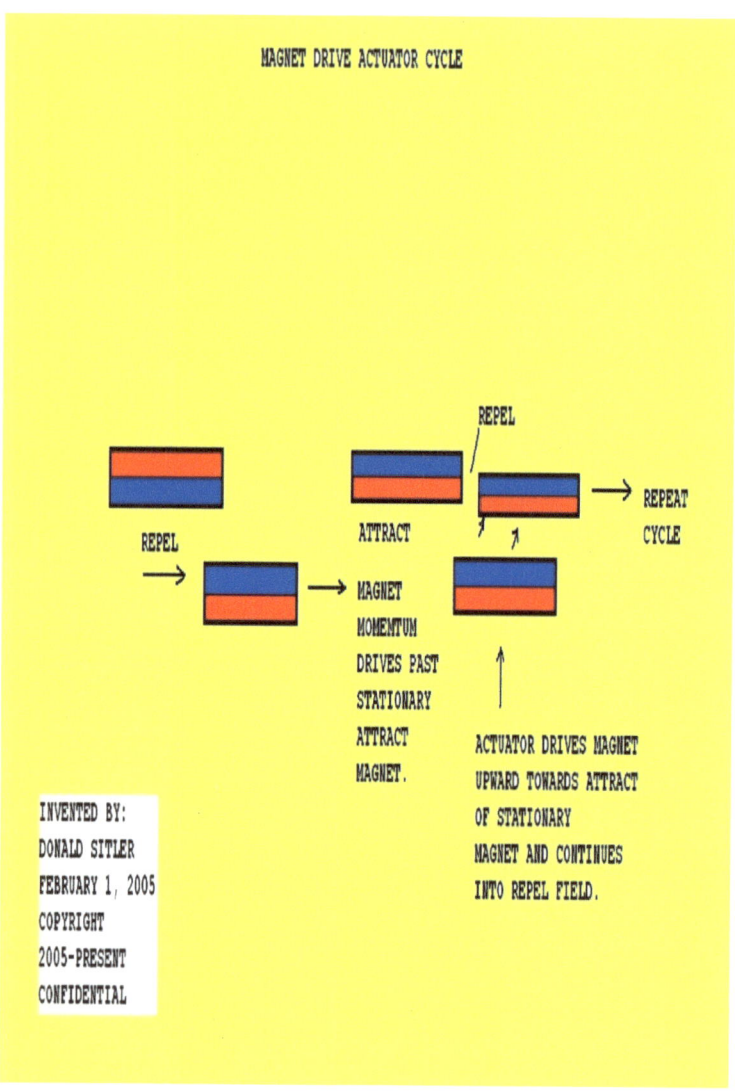

The magnetic field switch, cycle one to cycle two.

POWER ASSISTED MAGNET MOTOR

The illustration shows the field magnet as the lower magnet in multiple pictures as movement occurs. First, the field magnet is in repel with the first rotor magnet. As the rotor spins, the field magnet enters into attraction with the second rotor magnet. As momentum carries the rotor past the lock point of the second rotor magnet, the motor enters the "magnetic window", which allows the field magnet to move up. The magnetic field is weaker between the magnets at this point, which allows easier actuator movement, switching the magnetic fields that interact, resulting in a repulsion force between the second rotor magnet and the field magnet. This sets the stage for the second magnetic cycle to commence. These magnetic cycles repeat over and over as motor speed increases and accelerates.

Magnetic Collisions Avoided

The magnetic path of repel to attract in the magnetic cycle, prevents any occurrence of a hard magnetic collision from occurring, since only two magnets are in the rotor, and one field magnet for interaction. A managed collision occurs in the direction of rotation, aiding rotation at the field switch point. This collision acts as a magnetic ratchet, to keep motor direction to always move in the same direction. Since the collision is in the direction of rotation, it does not stress the pinning material in the magnet, so the magnet remains undamaged with use. The pinning material holds the magnetic particles in place in the magnet to maintain polar magnet strength. The pinning material prevents the magnetic particles in the magnet from moving toward each other in attraction, which would weaken the magnetic field that "shines" out from the permanent magnet. With magnetic collisions avoided, and magnetic cycles reset with each field switch, motor operation is smooth and will usually self start as long as the motor stops short of magnetic lock. When multiple rotors are used, it becomes less likely that the motor will not self start. With enough rotors, the motor could in theory, operate in total attraction mode, as other attraction fields from other rotors would provide torque to maintain rotation of the motor during magnetic field switching. More detail about "modes of operation" will be covered in future chapters.

CHAPTER 4: ADVANTAGES DISCOVERED

At first, when I came up with this design of magnet motor, I did not realize the advantages that this design afforded. I just knew that it worked and allowed for rotation of the motor. I was not set up to officially test the motor, nor did I understand how I should test it and what mathematical formulas to use to compute horse power and convert motor output into watts produced. As time went on, the motor developed from a single rotor that was electrically switched with a mechanical contact on the rotor with a wire for a brush. This circuit improved to an electronic mos-fet driven, optical switching circuit to drive the actuator up or down in time with the rotor.

POWER ASSISTED MAGNET MOTOR

The advantages began to show up when I went to a dual rotor motor. I used two actuators switched with one circuit. The actuators were put in series to reduce power consumption, and I reversed one of the coil connections so the power generated by the actuators would tend to cancel, reducing power consumption. The motor would still perform the magnetic switch while the actuators were in series, increasing the resistance of the coils, and reducing the amount of power required to run the motor by about 20 percent. I believe a third actuator could be used in series to further reduce power consumption. The second rotor aided rotation and smoothed motor performance. I found that motor advantages could really be increased by not having to power the actuators all the time. If the actuator could be held in a position with a magnet, the hold current would not be required to maintain the hold position while the actuator was in the up or in the down position. In essence, the actuator would only need to be pulsed during magnetic

window switch times, and not powered the rest of the time.
The hold magnet could maintain the position of up or
down without the need for holding power. This means that
two pulses per rotation would allow for the same motor
performance, without needlessly expending electrical power
on the actuators all the time. This could save 75 to 80
percent of the power consumption in the motor. The model
motor measured, using a prony brake and a digital scale, on
average .3 foot ounces of torque output at 1800 rpm.
That equates to over 4 watts of power output from the
motor. Motor power consumption was 6.25 watts on the
dual rotor model when powered 100 percent of the time.

As you can see from the previous chapters, the motor has
magnet drive during all but two points of rotation. I
estimated that the motor should only need to be powered
about 25 percent of the time, possibly only 20 percent of the
time. The motor while powered 100 percent of the time is
near 70 percent electrically efficient. This opens up superior

possibilities for efficiency compared to other motors,

estimated to achieve 200 percent, perhaps more on electrical

efficiency. If we can save 75 percent of power usage from

6.25 watts, the motor could consume 1.56 watts with about 4

watts output. Power input of 1.56 watts with 4 watts output

equals a potential 256 percent efficiency. That's enough to

get excited about. At saving 80 percent power usage, over

320 percent might be achieved.

I decided to write this book to explain the operation

of this motor so more people would gain and understand the

technology, increasing the chances of generating high interest

in this type of technology in the minds of readers. It is

apparent that this technology doesn't do any good if it just

sits in secret on the shelf. If no one can come to understand

the great importance and advantages of this technology

(lacking understanding of how it all works together), there

can be no path for this technology to see implementation and

development through Dynamaticmotors.

POWER ASSISTED MAGNET MOTOR

With these advantages, hopefully the reader can see that this motor appears to have the ability to extract the potential energy from permanent magnets, while operating in this motor. If efficiencies can be maintained (I believe they can be), this motor could have self running abilities with electrical power generation feeding back to the motor. This would allow running day or night, rain or shine. This motor is not well suited for start/stop applications. This motor would be better suited for a more constant running application, such as a fan, or turning a generator, as well as other non-start/stop applications that require high efficiency.

CHAPTER 5: MAGNETIC SPACING

During initial experimentation, while trying to find a way of magnetic propulsion, I made a model rotor with two magnets spaced evenly in the rotor. These magnets were alternating polarities that shined outward. I noticed when I took a magnet in hand, starting at the repel rotor magnet, it would thrust the rotor into a spin until the attraction fields took hold with the second rotor magnet. Rotation would continue for one half of a revolution until the field magnet (in hand) entered magnetic lock with the rotor magnet. No field switching by hand was needed to obtain a half revolution. I noticed that if I tried to move the magnet while in magnet lock position, it was very hard to move the field magnet.

23

However, if I held the magnet still until the rotor magnet passed the field magnet, then moved the field magnet to the magnetic switch position, the magnet moved much easier and instead of the magnets pushing sideways, the thrust was mostly in the same direction that the rotor was turning. Eureka! So I decided to make a fixed actuator that would be powered and timed to move in time with the rotor, at the ideal time for magnetic switch.

Once the model motor progressed to the form of an actual rotating motor, I could experiment with magnetic spacing of the field and rotor magnet. The field magnet needs to be vertically aligned with the magnetic fields in the rotor. Then the travel in the actuator needs to be positioned to reach both magnetic field positions. The field magnet needs to be positioned to reach equally, the low position rotor magnetic fields and the high position magnetic fields. This assures equal performance in both positions and also makes the field switch occur completely. Next, the spacing between the rotor and field magnet (side to side positioning) needs to be determined. If the spacing is too far, rotational torque will suffer. If the spacing side to side between the field magnet and the rotor magnet is too close, the torque is greatly increased, but so is the magnetic lock on attraction. There is a "sweet spot" that is between these two extremes, where the motor speed will be maximized. I adjusted the motor for maximized speed on my model. It seems to work quite well in this position for maximum efficiency. A closer in adjustment can be made, resulting in higher torque output, but it is at the cost of speed. It is an equation between spacing and rotational momentum. See the illustration, showing the relationship of the various magnets in the adjustment procedure.

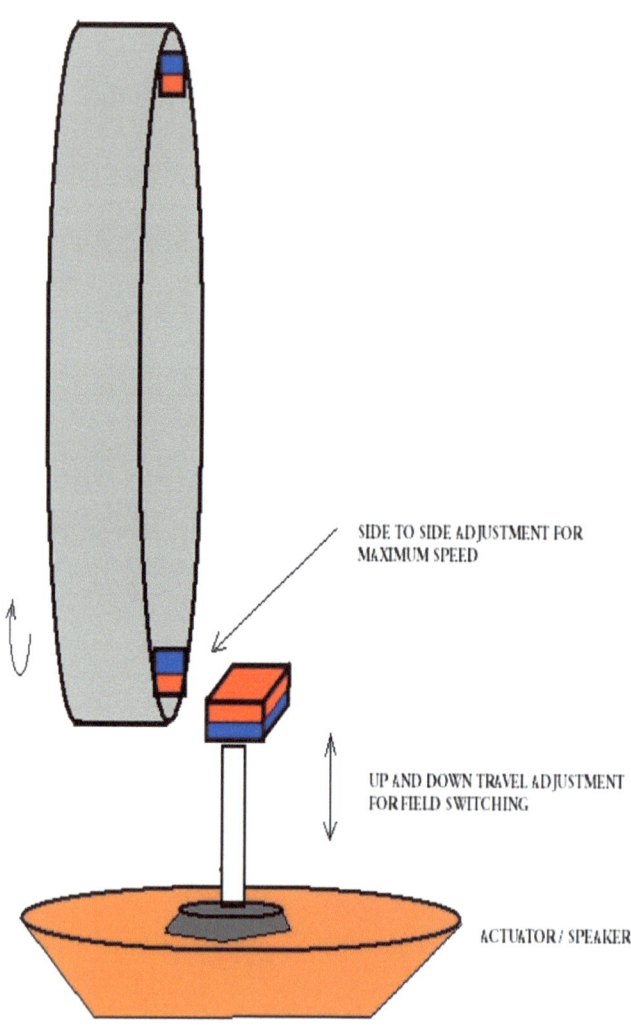

SIDE TO SIDE ADJUSTMENT FOR
MAXIMUM SPEED

UP AND DOWN TRAVEL ADJUSTMENT
FOR FIELD SWITCHING

ACTUATOR / SPEAKER

Magnetic spacing between field magnet and rotor magnets.

OPTICAL TIMING ADJUSTMENT

The optical timing assembly is located on the end of the axle.

Two optical sensors are located on the end of the axle. Only

one optical sensor circuit is used in the dual rotor motor,

which powers both actuators in series. A rubber grommet

was used with a groove in the outside that would allow a half

circle piece of cardboard to be installed in the groove that

would spin with the axle of the rotor. This rotor semi-circle

acts as an interrupter of the light source in the sensor

assembly, and is positioned in exact alignment with the rotor

magnets of rotor number one. When the light is interrupted,

the actuator would be powered in one position. When the

interrupter travels out of the sensor, allowing the light to be

sensed, the actuator will switch to the other position and hold

that position until the strobe is received to move again to the

other actuator position. Thus, the actuator is powered in one

direction or the other all the time. The actuator holds the

position until it is time for the magnetic field switch to occur

between the rotor magnet and the field magnet. The rotors

can be slid on the axle and the semi-circle interrupter can be moved for the actuator timing adjustment. The amount of advance timing determines how much of the repel field is utilized during the field switch. I have found it is useful to take advantage of the repel "push off" as it greatly increases speed. It also uses the field switch input power and uses it to add to the rotation of the motor, since that power will be used anyway to switch the motor's magnetic field.

5 volt sensor assembly for timing card input.

CHAPTER 6: THE ACTUATOR

While deciding what I would use for an actuator for field

switching between the field magnet and the rotor magnets,

several things had to be considered: cost, actuator travel,

speed, size, and magnet size of the actuator for efficiency. I

chose the Sony® 12 inch speaker for an actuator.

They were fairly inexpensive, and the cone allowed a fair

amount of travel for the magnetic field switch. The speaker

size was not ideal, but it would function as an actuator. The

12 inch size would slow actuator speed, but was good enough

for a model. The magnet size was robust, causing stronger

movement and greater cone travel. Due to the way the

motor is designed to switch the magnetic field, the rotor

magnet momentum carries the rotor past the actuator magnet and past the strength of the actuator's magnetic field. The rotor magnet is now positioned slightly past the magnetic field in the magnetic switching window. This momentum allowed the rotor magnets to pass the actuator's field strength, reducing the strength needed to move the actuator. This permitted less input power to be expended to perform the magnetic field switching during the magnetic window.

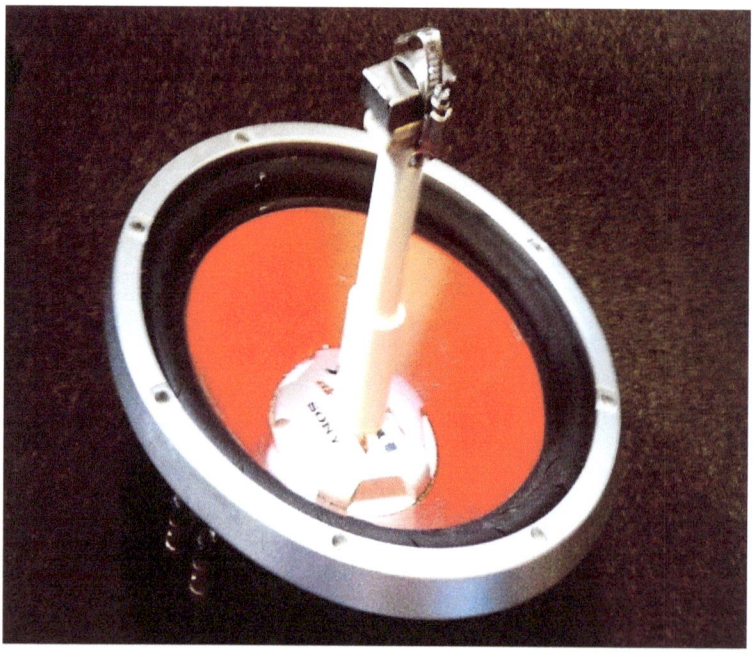

Sony® 12 inch speaker (XS-L1237) used as an actuator in the Power Assisted Magnet Motor.

On the center of the cone, a PVC ¾ inch pipe coupler was attached vertically using plastic pipe glue. Then an appropriate length of pipe was fitted inside of the coupler that was attached to the silver center of the speaker cone (dust cover). I used a stainless steel band to attach the field or stator magnet to the pipe stem. Two slots were cut into the pipe to slide the stainless steel band through the end of the pipe to firmly attach the magnet to the actuator pipe stem. While checking the travel of the magnet on the actuator, this indicated the length needed to make the pipe holding the magnet. The PVC coupler allowed for slight adjustment in height of the magnet, at the top of travel alignment and the bottom of travel alignment to the rotor magnetic path.

CONNECTION OF THE ACTUATORS

The actuators are connected in series in opposite phase. This causes the actuators to move in opposite directions when powered. The induced current in the actuators during operation will tend to cancel each other, reducing power input requirements and increase reaction time to changing direction. This also means that the second rotor will have its rotor magnets in reverse polarity of the first rotor. The second rotor will be switching in the opposite direction at the same time the first rotor switches. The actuators being connected in series, out of phase, will then connect to the H-Bridge switch final output stage. The function of the electronics in the next chapter will cover the function of this component as well as others.

CHAPTER 7: THE ELECTRONICS

The power assisted magnet motor consists of three

components that connect together to perform the needed

tasks to allow this motor to perform with precision timing.

The first portion is the sensor assembly. It consists of two 5

volt light sensors with light sensitive transistors acting as the

sensors. This assembly is mounted on the axle of the rotor.

A semi-circle shaped interrupter is mounted on the axle and

the leading and lagging edges of the semi-circle are aligned

with the magnetic window switching points of the magnets

mounted in the rotor. The light sensor will sense light when

the semi-circle interrupter is not blocking the light source in

the sensor, which represents the actuator positioned at one

extreme, until the sensor senses darkness. This would cause the actuator to move to the other position and hold until the sensor senses the next change. The actuator therefore moves from one extreme position to the other, within the magnetic window switching position.

The next component is the timing card. It uses the output of the light sensors as an input to the timing card. The timing card consists of a CMOS dual flip flop (4013), and a CMOS 555 clock chip for the (4013) chip. The flip flop acts as a logic level driver for the H-bridge output, which is the third component which drives the actuators. The H-bridge allows a separate voltage for the H-bridge switch portion, which permits a higher voltage source rather than the logic 5 volts for powering the load actuators. This is why a 5 volt regulator is employed on the timing card to maintain logic levels for stability, until the final stage of the H-bridge output. The parts list is shown with a list of all parts except the standard 5 volt light sensors.

DUAL ROTOR TIMING CARD PARTS LIST

PARTS:

2 - 1K Ohm Resistor 1/2 Watt
2 - 10K Ohm Resistor 1/2 Watt
2 - 4.7K Ohm Resistor 1/2 Watt

1 - 100mf 35v Capacitor Electrolitic
1 - .01 MF 35v or better disc Capacitor

1 - 14 Pin TI 6D4013BE Dual Flip Flop D Pos. - Edge CMOS
1 - 14 Pin Socket

1 - 3 Pin 7805A 5V Regulator

1 - 8 Pin TLC555CP CMOS Timer (Clock Source)
1 - 8 Pin Socket

2 - NMIH-0050 H - Bridge 5A (6A Surge) 5.3V - 40V (6V)

Parts list of the timing card.

POWER ASSISTED MAGNET MOTOR

SCHEMATIC DIAGRAM OF TIMING CARD

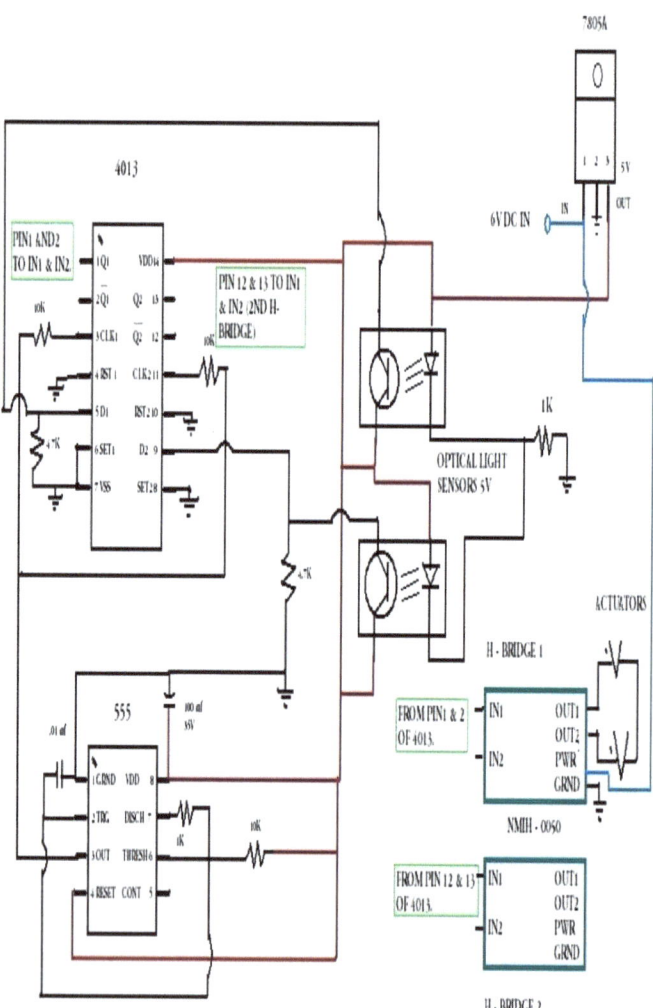

Schematic diagram of the timing card, including the light sensors and H-bridge components.

POWER ASSISTED MAGNET MOTOR

Here are a few notes about the timing card, the power consumption is very low due to the addition of a few resistors and the use of CMOS chips. Only the first output circuit is used on the dual rotor motor. The second circuit and H-bridge circuit can operate independently of the first circuit. It is triggered by the second light sensor. The 555 timer chip provides the clock for the 4013 dual flip flop. The capacitor on pin 1 is a .01 mf capacitor, which sets the frequency of the clock. The clock frequency is set high enough to avoid any limitation on the speed of the motor. The other capacitor is 100 mf, and is located between pin 1 and pin 8. This capacitor is there to provide stability to the 555 timer clock chip.

In the upper right corner of the schematic diagram, there is a 3 pin 7805 series pass regulator chip. It provides a stable +5 volts to all of the logic circuits. The power input is 6 volts DC to the regulator chip and the higher voltage (6 volts) is passed down to the H-bridge circuits to power the actuators.

POWER ASSISTED MAGNET MOTOR

Note that the phasing of the actuators is shown in the schematic with a small mark on the lead towards the H-bridge connection. The actuators are phased in reverse from each other. Therefore, the actuators will move in the opposite direction from each other when powered. This seems to reduce the power generated by each actuator, and reduce power consumption during operation and improve overall performance.

There are no heat sinks needed on any device except for the H-bridge chip (it comes with a heat sink, not that it is needed). The H-bridge is a NMIH – 0050. It can be configured several different ways. In this application, it uses two inputs, which come from the 4013 logic chip (pins 1 and 2). The inputs alter high and low inputs in time with the light sensor, which sense axle and magnet position at the axle. I am not sure that this H-bridge is still available. It is expensive, and can handle at least 5 amps, and up to 40 volts. When I made a decision on this chip I didn't want it to be inadequate, so it is very overrated. I would think that a

suitable replacement of 2.0 to 2.5 amps and up to 16-20 volts

would be fine, as long as it can handle the same inputs from

the 4013 dual flip flop chip.

Timing card that was used on the Power Assisted Magnet Motor.

POWER ASSISTED MAGNET MOTOR

The NMIH-0050 H-Bridge

Features:

- 5 A continuous, 6 A peak current
- Supply voltages from 5.3V up to 40V
- Terminal block for power / motor
- Onboard LEDs for motor operation/direction
- Onboard LED for motor supply
- Single row header for inputs
- Onboard LEDs for input indication
- Onboard LED for digital supply 5V
- Very low RDS ON - typically 200 mOhm @ 25 °C per switch
- Internal freewheeling diodes
- No crossover current
- Undervoltage lockout with hysteresis
- Overtemperature protection with hysteresis
- Output short circuit protected
- Error Flag signal indicator for fault conditions
- Onboard LED for Error Flag for fault detection
- CMOS/TTL compatible inputs with hysteresis
- Wide temperature range; - 40 °C < Tj < 150 °C
- Operating frequency up to 33 kHz

Benefits:

- Visual indication of status: power, inputs, and drive outputs
- Compact size
- Good thermal distribution
- Internally protected
- Simple connections
- Easy interface
- Easy mounting options w/brackets

The NMIH-0050 Full H-Bridge consists of a dual half h-bridge, screw terminal and pin connectors, single input inversion, and LED status indicators. It is usable for forward-reverse-stop control of many DC servomotors using TTL/CMOS compatible inputs to interface directly with most microcontrollers. It can also be used for phase control of steppers or brushless DC motors. With current handling to 5 A continuous, 6 A peak, it allows more flexibility in use with many small to mid range DC motors. Voltage supply range is from 5.3 to 40 V for the NMIH-0050. Switching characteristics allow PWM frequencies up to 33 KHz. Undervoltage protection, as well as Error Flag generation for shorted outputs, over temperature and over current are available for use.

NOTES:

CHAPTER 8: TWO DISCONTINUOUS FUNCTIONS CONNECTED

A discontinuous function is a line that is graphed that is not continuous and is equal to zero, when attempting to graph such a function or line. It has a break in it, like in the case of this magnet motor. In the picture illustration, we have a set of magnets in a linear fashion, or could be circular fashion. So, for this discussion we will treat it as linear. First, there is magnet number one followed by magnet number two. We have a line or function from magnet one to magnet two. The first magnet would be interacting with a third magnet which would be attached to an actuator. We have magnet number one in repel to the actuator magnet, and the force would be away from magnet one towards magnet two, which is in

attract.

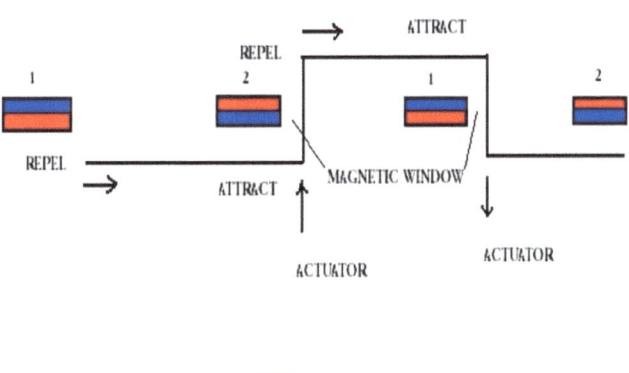

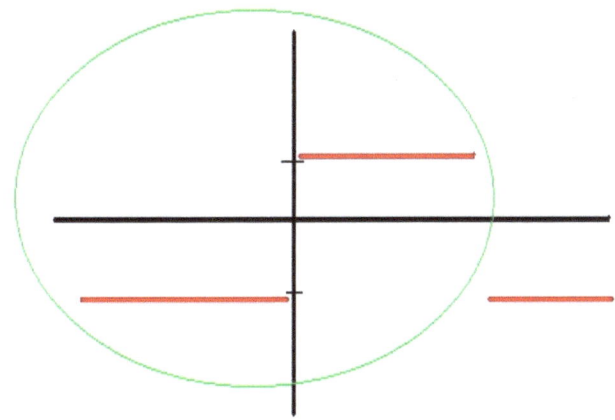

TWO DISCONTINUOUS LINES (FUNCTIONS)

POWER ASSISTED MAGNET MOTOR

When we travel from magnet one to magnet two, we have completed a straight line. We have the same line from magnet two to magnet one, when we use the even horizontal alignment with the magnets, which is the second line (the line with the magnets is shown above the magnets). This makes two lines that start and stop at two different levels in height. We are using the actuator to "connect" these two line functions that are actually the magnetic path of repel to attract. I have tried to show this in the diagram of a graph with the magnets above.

We know that the function of a pure magnetic motor path is equal to zero, mathematically showing that a magnet motor can not perform work. However, when we consider two discontinuous lines or functions, and connect them together using an actuator with a magnet attached, we now have two "connected" discontinuous lines, which when connected do not equal zero. We have a great deal of gain from these two segments of a magnetic path. This is why this motor actually

44

works and functions. I wanted to use somewhat of a mathematical comparison so that it makes another way of seeing that this motor can function, as long as we use an actuator to switch fields. It then becomes very important when and how we switch this field, to perform it in an efficient manner, utilizing less power than the strength of the magnets. This is accomplished by switching after or past the strength of the magnetic field, in the "magnetic window". This results in efficient magnetic field switching, and adds the power expended in the actuator motion, to the rotation of the motor by compressing the repel field in the magnetic window. This provides additional rotation to the motor, while at the same time accomplishing the magnetic field switch, connecting the two line functions, which are the operational magnetic field paths. In a circular arrangement, (instead of a linear arrangement) this would complete one full revolution of the motor.

CHAPTER 9: MODES OF OPERATION

The modes of operation can be better understood as an aspect of timing options within the magnetic window. For example, the timing of the magnetic field switch can be delayed to the point where it has passed the repel field. When this happens, the energy that was used in the actuator is not passed on and added to the motion of the motor. This results in a loss of momentum from the previous magnetic drive cycle. The motor at this point turns only due to attraction forces and the magnetic field switching to avoid magnetic lock. During experimentation while operating in this mode, the speed of the motor was much slower but continued to run. The advantage of this mode is that less actuator power is needed to perform the field switch, because

the actuator magnet does not encounter the repel field. It

may be useful in a multiple rotor and actuator arrangement,

with magnetic drive placed at regular intervals around the

rotor. This will provide magnetic drive around the rotor

from various magnets at the same time. They should be able

to operate on low power, while attraction forces provide

magnetic drive. A hybrid combination of using a small

amount of repel timing without increasing power usage

much, may increase performance. This would add some of

the input energy used in the magnetic switch, and use

this energy to add to motor rotation. Many different modes

of motor operation are possible in this configuration.

I call the attraction side of the magnet, the "free energy side".

The reason for this is that when you start with the rotor in

repel to the actuator magnet, then the rotor rotates to the

next rotor magnet in attraction. At this point, no input

power has been used to achieve one half revolution. But in

order to break the attraction lock potential, a field switch to repel is needed during the magnetic window. Input power on the actuator will be needed at this time to connect to the next magnetic cycle. When you encounter the repel field, input power is required to compress the repel field. The amount of torque that results from the compression of the repel field experiences a loss in rotational torque vs. the power used in the actuator. This action is not without loss because the torque produced is not exactly in the same direction as the motor is turning, but most of the repel torque still ends up aiding motor rotation. So it may be possible to create a motor that uses only attraction forces for magnetic drive, with minimal actuator input power for the purpose of breaking the attraction lock, not adding repel torque. This motor mode would yield an all magnet drive to rotate the motor. It would require multiple magnets and low power actuators.

The dual rotor model motor used a portion of repel field

capture during the magnetic window switch to increase speed and efficiency. This is due to the few number of magnets used in the motor, and the natural gap between rotor magnets.

CHAPTER 10: DYNAMATICMOTORS: INVENTION KNOWLEDGE

I have enjoyed the opportunity of engaging in creating and experimenting with these power assisted magnet motors and increasing my understanding of their operation. This has helped me to be able to improve my designs greatly. It has been fun to write about these machines and explain how they work to others with similar interests, because now is the time to share this knowledge and teach this new avenue of thinking. Schools and Colleges could offer this book as a resource to students that are in the process of learning how permanent magnet motors work with generally higher overall efficiency, than strictly electromagnetic motors. The strength of these magnetic motors should be improved for stronger commercial use.

I hope that this book generates strong interest with schools, developers and engineers that are interested in developing these new motors for commercial use. The reason I wrote this book is to provide a rich inside resource and theory of operation book, which will bring students, schools, engineers and interested readers up to speed, thinking outside of the box. This would result in focusing their attention on the many advantages that this new motor can provide. If the schools and engineers don't understand the real advantages of this new technology motor, they will never realize its value to advance this industry to the next level.

The use of a mechanical actuator is preferred compared to the use of a coil directly in the motor. This protects the motor gain in efficiency (due to the generator effect of a coil in a motor and the requirement to power the coil to the point

of equaling the strength of the magnetic field present in the

motor). Additionally, this permits non-powered

magnets of greater strength to interact with one another to

provide propulsion. This opens the possibility of higher

electrical efficiency than is possible with normal

electromagnetic motors. The actuator also controls the speed

of the motor while opening the door to higher levels of

efficiency. This is a reasonable compromise, that makes

possible this new level of efficiency, compared to other

electromagnetic motors.

The motor design includes a voice coil (as a moving part),

which could be considered a point of potential failure.

However, I think it is no longer a point of concern today

due to better design and materials, as well as better

engineering practices to reduce the likelihood of becoming a

point of failure. Much better magnetic holding voice coil

actuators can be used rather than a 12 inch speaker for this motor commercially. I see the future of these motors actually

being a clean power producer instead of a power consumer,

with proper actuator and motor design.

Thank you for purchasing this book and thereby supporting

my work. I hope I have opened some eyes and minds,

stirring strong interest with this knowledge, to the next level.

ABOUT THE AUTHOR

Don Sitler is married to Cindy Sitler, having two children. Don was the third born child, raised most of his early life in Lawrence, Kansas. Don self taught himself analog and digital electronics as a hobby and also learned from his Dad, who worked for Unisys in computer engineering. Don has also worked in computer electronic maintenance and computer support with Telex, Unisys, TSS, and IBM. Don enjoys experimenting and creating, while thinking outside of the box. Don attended college classes at Kansas University, and lives in Garnett, Kansas.

www.ingramcontent.com/pod-product-compliance
Lightning Source LLC
Chambersburg PA
CBHW040848180526
45159CB00001B/352